William Roberto Andrade da Silva
Luciano de Souza Cabral

SMART GUIDE

William Roberto Andrade da Silva
Luciano de Souza Cabral

SMART GUIDE

AN ACCESSIBILITY TOOL FOR PEOPLE WHO ARE BLIND OR HAVE LOW VISION

ScienciaScripts

Imprint

Any brand names and product names mentioned in this book are subject to trademark, brand or patent protection and are trademarks or registered trademarks of their respective holders. The use of brand names, product names, common names, trade names, product descriptions etc. even without a particular marking in this work is in no way to be construed to mean that such names may be regarded as unrestricted in respect of trademark and brand protection legislation and could thus be used by anyone.

Cover image: www.ingimage.com

This book is a translation from the original published under ISBN 978-613-9-68957-6.

Publisher:
Sciencia Scripts
is a trademark of
Dodo Books Indian Ocean Ltd. and OmniScriptum S.R.L publishing group

120 High Road, East Finchley, London, N2 9ED, United Kingdom
Str. Armeneasca 28/1, office 1, Chisinau MD-2012, Republic of Moldova, Europe
Printed at: see last page
ISBN: 978-620-8-33593-9

Dedication I would like to dedicate this final project to my mother,without her I wouldn't have got this far.

ACKNOWLEDGEMENTS

I would firstly like to thank all the teachers at the Federal Institute of Education, Science and Technology of Pernambuco IFPE, Jaboatao dos Guararapes campus. In particular, I would like to thank Professor Luciano de Souza Cabral for all the support and guidance I have received over the last year. I would also like to thank all my classmates, all of whom, without exception, have been crucial in getting me this far.

And last but not least, I would like to thank Mathias Cordeiro de Oliveira and Thiago José Amaral Oliveira da Silva, undergraduate students in Systems Analysis and Development at the Federal Institute of Education, Science and Technology of Pernambuco, Jaboatao dos Guararapes campus, who actively participated in the process of producing the simulated database, by recording videos within the internal premises of the campus, always in a very participative and proactive manner. Many thanks to everyone.

SUMMARY

This course conclusion work proposes the construction of fundamental bases for the creation of an accessibility tool, also known as assistive technology or adaptive technology, for people with visual impairments, whether blind or with low vision, providing these people with a means of facilitating social integration, especially with regard to their academic training, The aim is to help them move around the internal premises of the Federal Institute of Education, Science and Technology of Pernambuco, Jaboatao dos Guararapes campus. In order to fulfil this objective, it is necessary to integrate various types of knowledge, techniques and methods that facilitate accessibility for people with visual impairments: Orientation and mobility, audio description of scenarios, assistive technologies, working in an integrated way with the most up-to-date techniques of computer vision and natural language processing, these being subfields of artificial intelligence, finally, this work will be built on the basis of the Design Science Research methodology of research and development.

Keywords: audiodescription. database, artificial intelligence.

SUMMARY

1. INTRODUCTION

In the history of mankind, we can find various records of the existence of people with visual impairments, such as the imposition of penalties in the Code of Hammurabi. In the Bible, in the book of Exodus, walking men were mentioned, who walked with the help of staffs in their hands, among many other historical records. Human vision comprises a complex set of physiological processes including not only the capture of light reflected from objects by the human eye, but also an ordered set of brain processes for identifying and memorising the information present in images. (Instituto dos Cegos, 2023)

The main characteristic of visual impairment is the total or partial loss, whether congenital or acquired, of the visual senses. There are two main groups in the medical literature that share similarities and at the same time differences, and these differences will be listed below.

• Blindness is nothing more than the total loss of vision, or almost no ability to see at all. It is understood as the total absence of visual reflexes as well as the perception of shadows and the light reflected by objects or the environment.

• Another group is made up of people with low vision or also known as people with low vision, whose main characteristic is persistent impairment of vision, even after specialised medical treatment, with some remnant of light being captured by at least one eye.According to data from the World Health Organisation (WHO), around 36 million people in the world are totally blind and another 217 million have low vision. In Brazil, based on data collected in the 2010 census by the Brazilian Institute of Geography and Statistics (IBGE), a total of 528,634 people are unable to see, and another 6,056,654 people have low vision. (WHO, 2019)

Diseases such as glaucoma, diabetic retinopathy, optic nerve atrophy and retinitis pigmentosa are the main causes of blindness in the general population, occurring in the vast majority of cases in the adult life of these people. Orientation and Mobility (O&M) is an extremely relevant topic when it comes to promoting accessibility for people with visual impairments. It is an area of special education aimed at rehabilitating people who are blind or have low vision, using a wide range of techniques, methods and procedures to this end, working in harmony with their remaining senses. The remaining senses, such as smell, touch, hearing, vestibular perception and residual vision, are developed in order to develop skills capable of providing these people with an understanding of landmarks, the incidence of clues along the way, the presence of obstacles, among many other possibilities. In isolation, orientation is the ability to recognise the environment and establish one's position in relation to it. It means awareness of space, objects and one's own body, which develops through repeated **sensory-motor** experiences in the physical environment. Mobility in isolation is the physical ability to move around spaces in a determined and safe way, as independently and autonomously as possible. In order to address the countless social impacts that visual impairment can have on a person's life, the International Classification of Functioning, Disability and Health (ICF) emphasises the social aspects of visual impairment in its regulations, and proposes a mechanism for measuring the impact that the social and physical environment has on the functional capacities of people with disabilities. This means that disability can shift from the disabled person to the environment in which they live (ICF, 2001).

1.1 Assistive Technologies

Assistive technologies, or simply adaptive technologies, are resources whose main aim is to minimise the impact that their disabilities can have on their lives. In other words, assistive technologies aim to provide favourable conditions for the development and promotion of accessibility in different contexts, thereby increasing the **functional capacities of** people with disabilities, whatever they may be. In Brazil, the Technical Aids Committee (CAT) of the Secretariat for Human Rights of the Presidency of the Republic describes assistive technologies as. "Assistive technology is an area of scientific knowledge, with an interdisciplinary character, which encompasses products, resources, methodologies, strategies, practices and services that aim to promote the functionality, related to the activity and participation of people with disabilities, incapacity or reduced mobility, aiming at their autonomy, independence, quality of life and social inclusion" (CAT, 2000). Assistive technologies can therefore refer to a piece of equipment, item or product that is used to increase the functional capacities of people with disabilities.

1.2 Audio description

Audiodescription (AD) is an accessibility resource that helps visually impaired people to understand more about the various elements and events in their environment. Audiodescription of scenes takes place by transforming visual language into verbal language. As a linguistic mediation activity, it is a form of **intersemiotic translation** (converting visual information into textual or audio information, and vice versa). It can be used in a variety of contexts, the most common of which are films, cultural shows, art exhibitions, sporting events, lectures, conferences, etc. (Audiodescription, 2023).
The description of settings and events in spaces such as films, theatre shows, museum exhibitions, among others, follow some basic premises, which will be described below:

- 1° You must describe what you see, what or who appears.

- 2° Cite physical features, clothes, colours and other visual elements.

- 3° Delimit where the actions take place, as well as the time in which they take place.

- 4° The framework, prioritising elements of greater importance, according to the context.

In these stages, the use of clear, simple and objective language should be prioritised, thus avoiding subjective interpretations that could impair the understanding of the scenes and their events.

1.3 Artificial Intelligence

Artificial intelligence (AI) is the ability of machines to simulate human intelligence in order to solve problems that are too complex or repetitive. Natural language processing (NLP) is a branch of artificial intelligence responsible for bringing together machine learning and linguistics, and is also responsible for studying the various problems relating to the generation and understanding of human natural language by machines. (eveo, 2023) Speech technologies refer to the processes of producing and perceiving the sounds used in spoken language. There are basically two technologies in this regard: speech recognition tools and speech synthesis tools. Specifically, speech synthesis is an important element in the development of assistive technology tools. Speech recognition is the process by which a machine identifies spoken words (Text-To-Speech, 2008).Computer vision is the ability of machines to simulate human vision, using specific algorithms and appropriate image processing techniques. It enables the development of solutions and tools that are useful for people's lives in general, especially for the visually impaired. "Computer vision represents a body of knowledge that seeks to artificially model human vision in order to replicate its functions through the development of advanced software and hardware." (Bradski, 2008)

1.4 Delimitation of the Theme

This work is geographically limited to the internal premises of the Federal Institute of Education, Science and Technology of Pernambuco, specifically the campus based in the city of Jaboatao dos Guararapes - PE. The results obtained from the data collected during the recordings made at the campus and presented in this course conclusion were developed through simulations planned by the author.

1.5 Motivation and justification

One of the main justifications for this coursework is the many difficulties that people with visual impairments may face when travelling around the campus, whether they are completely blind or have low vision. It should be noted that the facilities on this campus are in relative compliance with the minimum criteria for promoting accessibility in federal educational institutions. However, it is possible to see that even two years after its inauguration, problems still persist in certain parts of the institution that could prevent potential students with special needs from attending the institution on a regular basis, jeopardising their autonomy and safety; one of these problems is exemplified in **Figure 01** below.

Figure 01 - Failures due to lack of proper maintenance of tiled floors.

Source: Own elaboration.

Description: **Figure 01** shows the flaws in the tactile flooring installed at the Jaboatao dos Guararapes campus, such as the lack of said flooring at certain points where it was installed.

As a result, it is worth pointing out that according to data from the Brazilian Institute of Geography and Statistics (IBGE), in the state of Pernambuco, 100,000 people are considered blind, and only 5% of these people attend public schools on a regular basis. This percentage is even lower when you consider the attendance rate of visually impaired students at technical and higher education institutions.

1.6 Materials and Methods

For this final project, an LG k22 mobile phone camera <https://www.tudocelular.com/LG/fichas-tecnicas/n6462/LG-K22.html> was used to capture the images by recording short videos. In terms of software tools, the open source computer vision library OpenCV, Yolo and the Flickr 8k database were used, as well as the tesseract tool for optical character recognition and the natural language processing library Speech_recognition and Google Text-to-Speech.

A Samsung Essentials e30 notebook with an i3 7th Generation processor and 8GB of RAM was used to run the codes that implement the libraries mentioned above. Some of the codes were implemented and adapted from the Google Colab platform.

1.7 Objectives and Contributions

Due to the great importance of accessibility, especially for people with visual impairments, the aim is to create the fundamental bases of study for the creation of an assistive technology aimed at this public. The general objectives will be presented below, followed by the specific objectives that have guided the development of this document, both at the level of theoretical foundation and practical application based on the tests carried out with the simulated database produced at the institution, as well as the objectives of the project.

their results, based on the proposed research and development methodology. In addition, the results will be discussed on the basis of advanced computer vision and natural language processing algorithms, which are available via links attached to this document.

1.7.1 General objectives

The main aim of this final project is to propose the analysis and integration of various types of knowledge capable of promoting accessibility for visually impaired people, namely Orientation and Mobility (OM), Assistive Technologies (AT) and Audio Description (AD), working in an integrated way with natural language processing tools and computer vision tools.

1.7.2 Specific objectives

In order to achieve the general objective of this work, the following specific objectives were defined:

• Test the operation of the OpenCV library and its HOG (Histograms of Oriented Gradients) methods, as well as the operation of Haar Cascades classifiers for detecting and tracking pedestrians and other detections in pre-recorded images and video streams.

• Measuring the detection and tracking capabilities of the Yolo computer vision library.

• Study and evaluate the results of the tests with the still image database, using the Flickr 8k dataset to create captions for still images.

• Evaluate the operation of the Tesseract library for optical character recognition.

1.7.3 Expected contributions

Among the contributions expected from the development of this final project is the possibility for visually impaired people to move around the internal premises of the Federal Institute with greater safety and autonomy, in the most efficient way possible, by detecting and tracking pedestrians, as well as identifying obstacles and avoiding possible collisions with them.

1.8 Work structure

Chapter 02 contains the theoretical framework as well as the related works pertinent to the themes of the research carried out, which are essential for a good understanding of the order in which these themes are structured, as follows.

- Visual impairment;

- Orientation and Mobility (OM);

- Assistive Technologies (AT);

- Audio description (AD);

- Natural Language Processing (NLP);

- Computer Vision;

Later, in **chapter 03**, we will work on the development methodology proposed for this final year's work, the aim of which is to seek a solution to one or more problems, based on an understanding of these problems, building and evaluating artefacts, allowing situations to be transformed in order to find ways of modifying them to more satisfactory conditions.

Chapter 04 will present the analysis of the results, based on the tests carried out using the simulated database produced at the Federal Institute, in line with the technologies described in **Chapter 03** Finally, **Chapter 05** will present the final considerations on the development of this course conclusion work.

2. THEORETICAL FRAMEWORK AND RELATED WORKS

In this chapter, the main concepts and terms will be covered, as well as the theoretical basis for the themes proposed in this final course work. Firstly, the main aspects of human vision and the numerous physical and social problems caused by visual impairment will be described.

This will be followed by a discussion of the different ways of using orientation and mobility (OM) techniques and methods for people with visual impairments. Next, concepts relating to the main assistive technologies available on the market, whose main objective is to promote accessibility in public places for people who are blind or have low vision, will be explained. Later on, the main points relating to audio description of scenes will be described in order to make it easier for visually impaired people to understand the environment around them.

This will be followed by a discussion of the main concepts relating to natural language processing, as well as its different current use cases. Finally, the operation of the computer vision technologies discussed in this document will be described theoretically, in the order in which they are presented.

* OpenCV library for pedestrian detection and tracking.

* YOLO library for pedestrian tracking.

* Flickr 8k database for creating image captions.

* Tesseract library for optical character recognition.

2.1 Related work

The project entitled "USE OF COMPUTATIONAL VISION AND FEATURE DETECTION TO AID NAVIGATION OF VISUALLY DISABLED PEOPLE IN INTERIOR ENVIRONMENTS" (UFFS,2016), presented at the VI Scientific and Technological Initiation Conference held from 17 to 18 October 2016 at the Federal University of the Southern Frontier (UFFS).

- Campus Chapecô, proposed the use of information technology to assist the navigation for visually impaired people, proposing the construction of a low-cost object recognition system.

To this end, an application would be developed that would be installed on the user's smartphone, where the user would make a recording with the device's camera of the environment in which they are, and by means of sound signals, receive the results of the detections of the objects found, such as doors, stairs, lifts, among others, these detections would be possible through previously defined computer vision techniques.

With this in mind, it was decided to use the OpenCV computer vision library, where three methods were evaluated, which were considered pertinent to the project's objectives: Features Detection, Template Matching and HIstogram Comparison. Due to the low degree of assertiveness of the Template Matching method, it was decided to discard it before running the tests. The results of the tests are shown in the table below.

Table 03 - Comparison of results for choosing the most efficient method.

Método	Resultado	Objetos (72)	Percentual
Features Detection	Verdadeiro Positivo	36	50
	Verdadeiro Negativo	33	45,8
	Falso Positivo	3	4,2
	Falso Negativo	0	0
Histogram Comparison	Verdadeiro Positivo	28	38,9
	Verdadeiro Negativo	21	29,2
	Falso Positivo	23	31,9
	Falso Negativo	0	0

Source: UFFS (Adapted).

Description: **Table 03** contains the results of the tests carried out with the Features Detection and Histogram Comparison methods.

2.2 VISUAL IMPAIRMENT

Human beings are visual in their actions and behaviour, and because of this, a large part of the information present in the environment is captured by the eyes. Seeing is a highly complex physiological process that involves numerous structures in the human body. Light first passes through the eyes, then it is captured by the retina and sent to the brain, via the retina. through the optic nerve, and finally, the neurons are responsible for processing this information. (Flavio, 2019) According to data from the World Report on Disability (the first document to comprehensively address the needs and barriers that these people face in their daily lives), released in 2010, every 5 seconds 1 person becomes blind in the world, and of the total number of new cases of blindness, close to 90% of these cases occur in underdeveloped or emerging countries (WHO, 2011).Visual impairment is characterised as the total or partial loss of sight, whether due to congenital or acquired disease. Eye accidents represent a significant number of factors that can lead to new cases of total or partial loss of vision, resulting from injuries caused by direct contact with objects or contact with chemical substances such as household cleaning products, or even direct contact with flammable chemical products such as combustibles and corrosive chemical products such as acids, among other possibilities.

2.2.1 Blindness

Being the total inability to see, blindness can be of three types: congenital (it occurs between 0 and 1 year of age); early (it occurs between 1 and 3 years of age); or acquired (it occurs after 3 years of age). Because of this, a congenitally blind person, due to the absence or reduction of a visual reference (mental image), has an intellectualised representation of the world (environments, volumes, perspectives, colours, etc). (Bernardes, 2016)

Figure 02 - The main causes of blindness in the world.

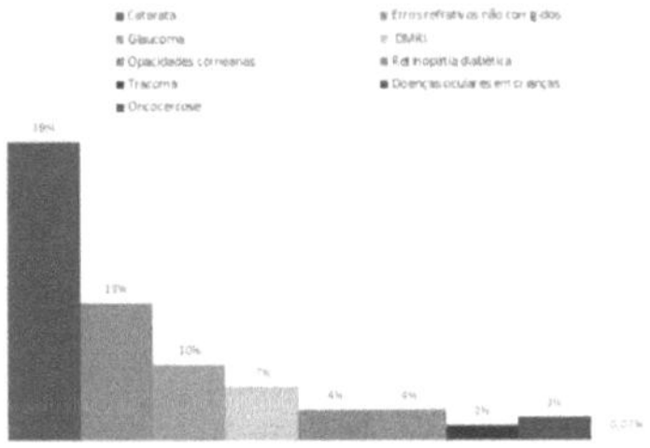

Source: (Saude Plena, 2019) (adapted) .

Description: Figure 02 shows a column graph with percentage values indicating the main causes of blindness in the world, according to data collected by the Brazilian Council of Ophthalmology in 2010.

Another important point worth emphasising concerns cases of partial or total blindness, which arise as a result of accidents with a strong impact on the head. However, the determining factors that can lead to blindness or low vision in the adult population are diseases such as glaucoma, diabetic retinopathy, optic nerve atrophy, retinitis pigmentosa and age-related molecular degeneration (AMD). Among children, the main causes are diseases such as congenital glaucoma, retinopathy of prematurity, congenital cataracts and congenital ocular toxoplasmosis.

2.2.2 Low vision

People with low vision or low vision have a reduction in their ability to see, even after correcting common refractive errors, the most common types being myopia, hyperopia, presbyopia and astigmatism, which can end up limiting their functional capacities. The main problems related to low vision are: low visual acuity, difficulties seeing near and/or far, reduced field of vision, as well as problems with contrast vision, among others. The use of any visual aid can help a person with low vision to perform many tasks, such as

- Mobility and getting around.

- Coordination of movements.

- Learning through reading and writing.

- In contact and relationship with the environment.

- On the constitution and organisation of space.

2.3 ORIENTATION AND MOBILITY (OM)

Promoting accessibility in public places for all individuals with physical limitations to move around safely and efficiently is the responsibility of all public administrations, whether at municipal, state or federal level. It is also the responsibility of private entities to implement in their spaces all the standards described in **Law No. 10.098 of 19 December 2000,** which establishes general rules and basic criteria for promoting accessibility for people with disabilities or reduced mobility (Felippe, 2001). "Orientation and Mobility is nothing more than a discipline that aims to help visually impaired people develop or establish the ability to move independently, efficiently and safely through spaces, to meet their own needs." (Felippe, 2001)

Because of this, there are various techniques that are worked on and taught by professionals specialising in the study and teaching of orientation and mobility: The OM technique which takes place, with the support of a sighted guide (people with sight, this can be a relative, a friend or even a stranger), another would be, through the use of a long cane, this being, the one that represents the greatest factor of autonomy and social integration, for people with visual impairment, and last but not least the technique that uses a guide dog, being an animal, properly trained to perform this task.

Table 01 - The three main techniques used by OM.

Technique	Description
Clairvoyant guide	Any person who, after prior acceptance by the disabled person, will help them move around in unfamiliar places. It is fundamentally characterised as a dependent form of locomotion.
Long cane	Capable of providing the visually impaired with crucial information about the environment using techniques such as: scanning, object detection, passing through doors, diagonal tracking, going up and down stairs, touch techniques, touch and swipe techniques, swipe techniques, touch tracking, three-point tracking, etc.
guide dog	A trained animal, with the aim of offering assistance and guidance during journeys, anticipating obstacles and, consequently, minimising potential accidents.

Source: Own elaboration.

Description: Table 01 contains the row descriptions of the three main orientation and mobility techniques that are widely used today.

One of the main responsibilities of the professional specialising in teaching orientation and mobility is to draw up a work plan according to the functional abilities of each student. The training is divided into several stages and, during these stages, some fundamental aspects of O&M teaching will be addressed.

• Cognitive aspects, which concern the segmentation of concepts, the nature and functions of objects, abstraction and the solution of various problems.

• Psychomotor aspects that provide O&M students with experiences in developing their perceptual abilities and basic movements.

• Emotional aspects, aimed at helping the student to acquire greater self-confidence, increased self-esteem and motivation.

Another relevant point for teaching O&M concerns the use of the remaining senses as a means of facilitating mobility, such as recognising sound signals common to a particular region, or even even familiarisation with certain odours characteristic of a place, constituting an important element in the OM teaching-learning process.

2.3.1 Clairvoyant guide

Important notes: The O&M technique that takes place with the help of a Guide can only take place with the prior consent of the visually impaired person. As for the O&M technique with the help of a guide, it is extremely important that the animal is never distracted by situations that jeopardise the performance of its activities.

2.3.2 Long cane

A tool that facilitates accessibility for the visually impaired, the long cane works as a kind of "extension of one's own body", its main purpose being to anticipate direct contact with possible obstacles and detours along the way, while also being able to provide the user with an empirical perception of a direction to follow, i.e. a path free of obstacles to their full mobility.

Through the use of the long cane, it is possible to benefit from another accessibility resource, which is extremely important for OM teaching. This resource is installed in both public and private places, and is called **tactile flooring**. They can be found in metro stations, bus terminals, streets, squares and various other places, as well as in private commercial establishments, shops, shopping centres, among others. This technology is present in the internal premises of the Federal Institute, Jaboatao dos Guararapes campus (Bittencourt, 2006).

They are also known as podotate floors and there are two fundamental types, which are physically similar to each other, but have different purposes that complement each other and should always be used together. **Table 02** below shows the fundamental difference between the two types of tactile flooring.

Table 02 - Distinction between the two types of flooring.

Types	Description
Directional tactile flooring	Its function is to indicate a direction or path, free of obstacles, to be followed. It is formed exclusively by lines in high relief on its surface.
Tactile warning floor	Its purpose is to indicate obstacles ahead, such as stairs, lift entrances, pillars and so on.

Source: Own elaboration.

Description: **Table 02** basically describes the purpose for which each type of tactile floor was created.

The set of techniques, tools and procedures that act as facilitators of accessibility for visually impaired people end up resulting in another concept, of extreme importance for the study and teaching of OM, which concerns the existence of an imaginary "Guide Line", referring to the perception of an indication of a direction to be followed when travelling.

The Jaboatao dos Guararapes campus has this important accessibility resource at its disposal. The physical form of the tactile floors is based on points of high relief, making it possible to feel the presence of this special floor by touch, with the aid of a long cane or with your own feet. The two types of tactile floors are shown in **Figure 03** below.

Figure 03 - The two types of podotate flooring available on the market.

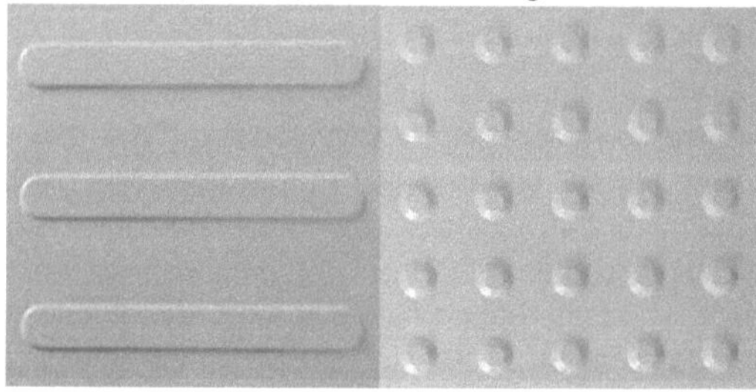

Source: (safe park signalling, 2023) (adapted).

Description: **Figure 03** shows the two types of tactile flooring available on the market, the first being directional tactile flooring, with spare lines that can be easily felt by anyone who stands on it. The second is the warning tactile floor or the popularly known "ball floor", which, as the name suggests, has the function of alerting the visually impaired to the presence of obstacles in front of them or even changes in direction.

2.3.2.1 Scanning Technique

This technique gives visually impaired people an almost immediate and comprehensive exploration of the terrain in the area close to their body.

2.3.3 Self-protection

The term self-protection encompasses a range of procedures which, if applied correctly, enable visually impaired people to be better protected when travelling. These procedures are carried out with the help of the body itself, using the body segments (upper and lower limbs, trunk, head), establishing positional and relational relationships, and self-protection is common to the other OM techniques mentioned in this document, such as the sighted guide, the long cane and the guide-chain.

2.3.3.1 Familialisation

It provides visually impaired people with an important means of studying their surroundings safely and efficiently in diverse and unfamiliar environments.

2.4 ASSISTIVE TECHNOLOGIES (TA)

Generally speaking, assistive technologies represent a set of items, parts or equipment, the use of which is intended to increase people's mobility. functional capacities of people with disabilities. Adaptive technologies refer to items that have been developed specifically for use by people with disabilities (Assistive Technology and Education, 2023).
Assistive technologies are characterised by their transdisciplinary nature, which means that they encompass professionals from various fields of knowledge, some of which are worth mentioning:
- Education.
- Psychology.
- Engineering.
- Design.
- Physiotherapy.
- Technicians and professionals from other areas.

With regard to the existing regulations that deal with this issue, the Americans with Disabilities Act (ADA), which is a law created in 1990 in the United States of America, deals with the civil rights of people with disabilities, prohibiting any form of discrimination, considering the list of assistive technologies specialised in helping people with visual impairments, whether blind or with low vision (ADA, 2023).

Since its creation more than a century ago, the Braille language has been a tactile reading and writing system that to this day represents a major milestone in the quest for social integration for people with disabilities and all those who work towards this end, especially with regard to the inclusion of these people in basic education through to higher education, as well as their inclusion in the labour market. The Braille system is basically made up of a set of dots in high relief, structured in units called Braille cells, which can be easily felt through contact with the fingers of the hand, usually read with the index finger. The Braille alphabet is shown in **Figure 04** below. (Colégio gtm, 2023)

Figure 04 - Representation of the Braille alphabet.

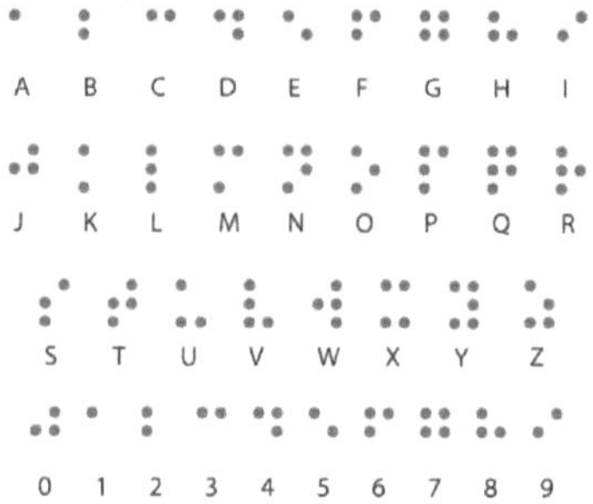

A B C D E F G H I

J K L M N O P Q R

S T U V W X Y Z

0 1 2 3 4 5 6 7 8 9

Source: (freepik, 2023) (adapted).

Description: **Figure 04** contains all the symbols that make up the Braille alphabet, each cell has a corresponding letter, where they are arranged from A to Z, and numbered 0 to 9 in that order.

2.4.1 Brazilian legislation

On 16 November 2006, under the competence of the Special Secretariat for Human Rights of the Presidency of the Republic - SEDH/PR, by means of Ministerial Order No. 142, the Technical Aids Committee - CAT was set up in order to bring together experts and representatives of government bodies around a common work agenda.

"For people without disabilities, technology makes things easier For people with disabilities, technology makes things possible." (RADABAUGH, 1993)

2.5 AUDIO DESCRIPTION (AD)

As an important accessibility resource, audio description of scenery broadens the understanding of visually impaired people about the environment around them and its events, whether at cultural events, theatre plays, films and art exhibitions, among many other possible examples, this information can be used in a variety of ways. They reach the user of this resource by means of sound signals, usually spoken by an individual, representing an additional narrative track, which can be reproduced in pre-recorded form or generated in real time as the facts in the examples cited above develop. (Motta, 2010)

2.5.1 General Principles

Audiodescription basically consists of a clear and objective description of all the elements visually present in the environment, allowing people to receive the information contained in images, both static and dynamic (videos). The professional audio-describer must describe:

- Information about the environment.

- Facial and body expressions.

- Changes in time and space.

In general, the audio description should be clear and objective, and should not contain inferences or interpretations that might jeopardise the understanding of what is being described.

2.5.2 History and Brazilian Legislation

The internationally recognised symbol for audio description consists of the letters A and D, with three brackets to the right of the letter D, alluding to sound waves propagating. In the United States, audio description emerged gradually in 1975, while in Brazil, audio description began in 2003.

Table 04 - Brazilian legislation on audio description.

Legislação aplicada à Audiodescrição	
Lei nº 10.098, de 19 de dezembro de 2000	Estabelece normas gerais e critérios básicos para a promoção da acessibilidade das pessoas portadoras de deficiência ou com mobilidade reduzida, e dá outras providências.
Decreto Federal nº 5.296, de 2 de dezembro de 2004	Regulamenta as Leis nos 10.048, de 8 de novembro de 2000, que dá prioridade de atendimento às pessoas que especifica, e 10.098, de 19 de dezembro de 2000, que estabelece normas gerais e critérios básicos para a promoção da acessibilidade das pessoas portadoras de deficiência ou com mobilidade reduzida, e dá outras providências.
Portaria nº 310, de 27 de junho de 2006, do Ministério das Comunicações	Aprova a Norma Complementar nº 01/2006 - Recursos de acessibilidade, para pessoas com deficiência, na programação veiculada nos serviços de radiodifusão de sons e imagens e de retransmissão de televisão.
Decreto nº 6.949, de 25 de agosto de 2009	Promulga a Convenção Internacional sobre os Direitos das Pessoas com Deficiência e seu Protocolo Facultativo, assinados em Nova York, em 30 de março de 2007.
Portaria nº 188, de 24 de março de 2010	Altera o subitem 3.3 e o item 7 da Norma Complementar nº 01/2006 – Recursos de acessibilidade, para pessoas com deficiência, na programação veiculada nos serviços de radiodifusão de sons e imagens e de retransmissão de televisão, aprovada pela Portaria nº 310, de 27 de junho de 2006.
Portaria nº 312, de 26 de junho de 2012 (DOU de 29/06/12, página 63)	Altera texto do item 7.1 da Norma Complementar nº 1/2006, estabelecendo valor mínimo de horas para veiculação obrigatória do recurso de legenda oculta para emissoras do serviço de sons e imagens e de retransmissão de televisão.
Instrução Normativa Ancine nº 116, de 18 de dezembro de 2014	Dispõe sobre as normas gerais e critérios básicos de acessibilidade a serem observados por projetos audiovisuais financiados com recursos públicos federais geridos pela Ancine; altera as Instruções Normativas nº 22/03, 44/05, 61/07 e 80/08, e dá outras providências.
Lei nº 13.146, de 6 de julho de 2015	Institui a Lei Brasileira de Inclusão da Pessoa com Deficiência (Estatuto da Pessoa com Deficiência).

Source: (Enap, 2023) (Adapted).

Description: **Table 04** shows the main legal instruments that deal with audio description in Brazil.

2.5.3 Simultaneous Audio Description (ADS)

In simultaneous audio description, all the phases of the audio description take place at the same time as the events. This usually happens when the audio-describer has no prior knowledge of the events. It is advisable to use specific verbs to indicate how the actions are carried out, e.g. run, jump, crouch, etc. During the audio description of a character's physical attributes, the following sequence is recommended: gender, age group, ethnicity, feather colour, height, physical complexion, eyes, hair and other relevant characteristics.

2.6 NATURAL LANGUAGE PROCESSING

The main aim of Natural Language Processing is to combine machine learning with the elements of human language (linguistics). In other words, it aims to understand human language and enable machines to understand and respond intelligently and comprehensibly within that language. In general, the phases of natural language processing segment human language into smaller, essential parts, making it possible to understand and relate these smaller parts in order to create a greater meaning for these parts. (Jéssica Rodrigues, 2017)

Figure 05 - Fields of study based on Natural Language Processing.

Source: (Jones Granatyr, 2020) (adapted).

Description: **Figure 05** contains a set of 7 images of the main fields of study in natural language processing, in the order they were arranged in **Figure 05**: speech transcription, neural machine translation, chatbots (intelligent chat), Q&A, text summarisation, image captions and finally video captions.

Nowadays, systems based on Natural Language Processing are developed based on premises that take into account not only the clear need to understand the literal meaning of the words in a text or speech, but also seek to understand how to interpret texts in a satisfactory way, being able to analyse the various intentions of the authors, as well as the reader, and being able to detect possible feelings expressed in these texts in a hidden or difficult-to-understand way. (Leandro Kovacs, 2021)

2.6.1 Voice Recognition

Speech recognition systems (ASR) can be separated into different types, which determine the way in which speech recognition is carried out (Caleb et al, 2008):

* Single words;

* Connected words;

* Continuous speech;

* Spontaneous speech;

* Voice verification / identification.

2.6.2 Voice Synthesis

Voice synthesis is the artificial creation of the human voice. It refers to a system that uses a synthesiser for this purpose and can be implemented in software or hardware. It can be created by merging pre-recorded speech excerpts stored in a database specifically for this purpose, or by simulating the human voice synthetically.

Figure 06 - Flow for processing texts within TTS.

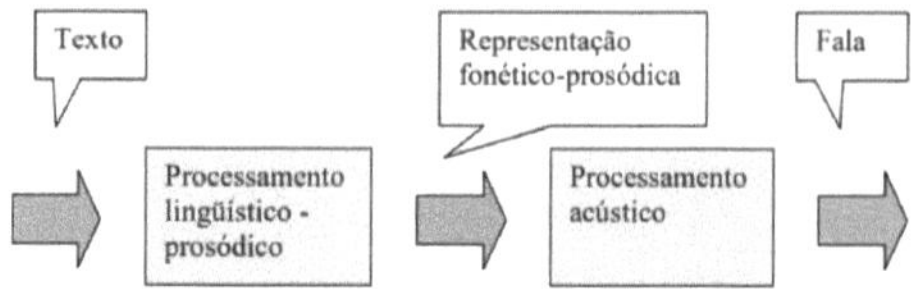

Source: (Caleb et al, 2008) (Adapted).

Description: **Figure 06** shows the flow of text processing to convert it into speech using speech synthesis.

2.7 COMPUTER VISION

Computer vision is the ability of machines to process images captured from the real world, enabling them to interpret the information contained in these images in a logical and cohesive way. Even at a stage when computer vision can still be considered in its infancy, it has made it possible for applications based on this knowledge to appear more and more frequently.

Applications developed for military, commercial, public security, business or the development of more agile and efficient production processes. These advances have occurred to a greater extent, especially in recent years, and it is worth highlighting the two main companies that are leaders in this segment, Testa and Google's Waymo, both of which are focused on developing and improving the embedded systems present in autonomous cars (Bradski, 2008).

Among the various types of tools and/or functionalities that have been developed in recent years based on computer vision, a few stand out:

- Detection of cancerous tumours based on the analysis of CT scans.
- Qualitative analyses of products in the production processes of large industries.
- Analysing lung x-rays based on the covid-19 case.
- Face detection and recognition.
- Pedestrian and object tracking.
- Vehicle automation.

Figure 07 - Areas of knowledge that benefit from computer vision.

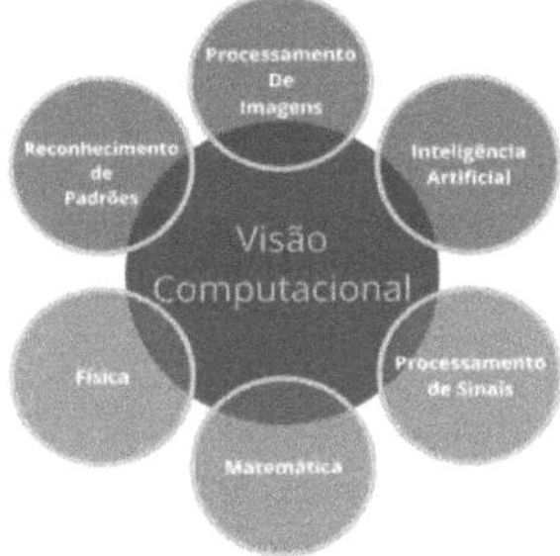

Source: (PrograMaria, 2020) (adapted) .

Description: **Figure 07** shows a larger, centralised circle that represents the field of computer vision, intersecting with this larger circle are other fields of knowledge, such as mathematics, physics, artificial intelligence, among others.

2.7.1 OpenCV

OpenCV is an open source library for developing computer vision-based applications. It is supported in programming languages such as Python, C++ and Java. OpenCV currently has more than 500 functions in its collection, which are easy to implement, making it one of the most widely used libraries for developing solutions in this segment. It is widely used for image processing and image pre-processing, among many other use cases. (OpenCV, 2023)

Among these more than 500 different methods, the thresholding method has the function of segmenting one or more characteristics present in an image, separating the object of interest, which is usually in the foreground, from the background of this same image.

Figure 08 - Using the thresholding method to segment objects.

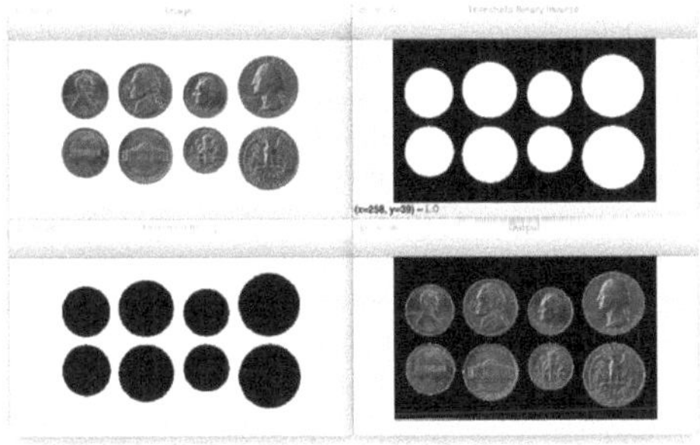

Source: (Adrian Rosebrock, 2021) (adapted) .

Description: **Figure 08** shows the different types of results obtained by using the thresholding method on an image. In the image in question, there are coins of different sizes and colours, segmenting and separating these coins in relation to the background of this image.

2.7.1.1 Histogram of oriented gradients (HOG)

Before dealing with the main concepts relating to the HOG method, it is necessary to briefly contextualise what visual descriptors are for computer vision. The purpose of visual descriptors is to describe the visual resources present in an image, describing its elementary characteristics, such as: the distribution of colours, the presence of specific geometric shapes, the existence of different types of textures, among many other characteristics. (SILVA, 2018)

This makes it possible to extract specific characteristics of one or more people in an image, such as: head, arms, legs, hands, skin colour, eye colour, among others. These characteristics can be used to build and train a new machine learning model in order to obtain better results for the detentions, improving the results obtained with each new training run.

2.7.1.2 Haar Cascades sorters

Haar Cascades classifiers represent an approach based on machine learning, where positive and negative images are used to train the classifier. The fundamental differences between positive and negative images are described in **Table 05** below. (Arnaldo Gualberto, 2018)

Table 05 - Description of the different types of images used by the Haar Cascades classifier.

Positive images	Negative Images
Contains the images you want the classifier to identify.	They represent everything else, i.e. images that do not contain the object/item of interest.

Source: Own elaboration.

Description: **Table 05** contains a description of the differences between positive and negative images when training the Haar Cascades classifier.

In their article entitled, "Rapid Object Detection using a Boosted Cascade of Simple Features" (Paul Viola and Michael Jones, 2001), they proposed the creation of an algorithm capable of detecting objects present in images, regardless of their location or scale, while still being able to detect objects in pre-recorded video streams or in real time. **Figure 09** below shows how this algorithm works on an image (Silva, 2018).

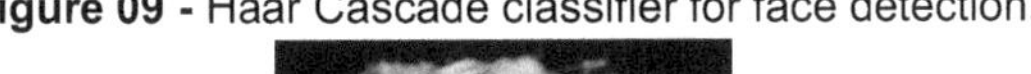

Figure 09 - Haar Cascade classifier for face detection.

Source: (Adrian Rosebrock, 2021) (adapted).

Description: **Figure 09** shows how the algorithm behind the Haar Cascades classifier runs through the image in order to detect the object/item of interest.
The Haar Cascade can be used to detect various types of objects, however, it is necessary to have a file with the extension ".xml", duly trained with the object you want to detect.

2.7.2 You Only Look Once (YOLO)

Developed for object detection and tracking based on single-shot convolutional neural networks, YOLO is an abbreviation for "you only look once". It was built and is maintained on an architecture known as Darknet. YOLOv5 (version number five, which is the most current) represents the first of the YOLO models, developed in the PyTorch framework <https://pytorch.org/docs/stable/index.html>, being considerably lighter and easier to use. (Joseph, 2015)

Figure 10 - Object tracking with YOLO.

Source: (Jobu, 2023) (adapted).

Description: Figure 10 shows how different types of objects are tracked in an image using YOLO. The image in question is in "gif" format, containing a man riding a bicycle, with a car in the background, these three distinct objects are detected simultaneously.

YOLOv5 works at a significantly higher speed compared to other contemporary algorithms for object detection and tracking, running at 45 FPS (forty-five frames per second, representing the number of images displayed on the screen in just one second).

2.7.3 Image captions

By using the dataset, it is possible to assign an appropriate caption to an image, containing the main characteristics of the image in question. A tool developed within this concept is Flickr, which is available free of charge on the Kaggle platform (Platform for Datascience Competitions) and is capable of working satisfactorily on computers with reduced processing power. (HSANKESARA, 2018)

2.7.3.1 Flickr 8k

Flickr 8k is a database capable of assigning appropriate captions to images. Its database contains 8000 images, divided into 3 sets. The first is the main training set containing 6000 images of reference. Another development-specific set containing a total of 1,000 images, and finally a test set made up of another 1,000 images. Each of these images has five different captions assigned to it to describe it (HSANKESARA, 2018).

An important point worth noting in relation to the creation of this database concerns the way it was selected, given that the images tend not to contain people or places known to the general public, and these images were selected manually in order to portray a greater variety of scenarios and situations.

With regard to the Flickr 8k numbers, it should be noted that since its last major update, which took place in 2020, 104 direct contributions have been made to the database. The total number of downloads since its creation exceeds 20,000.

Still on the subject of the data set and its formation, the two figures presented next in this chapter, **Figure 11** and **Figure 12**, show respectively how the five captions were distributed to each reference image, then the different scenarios and situations represented in the images in the database.

Figure 11 - Arrangement of the 5 captions assigned to each of the 8,000 images.

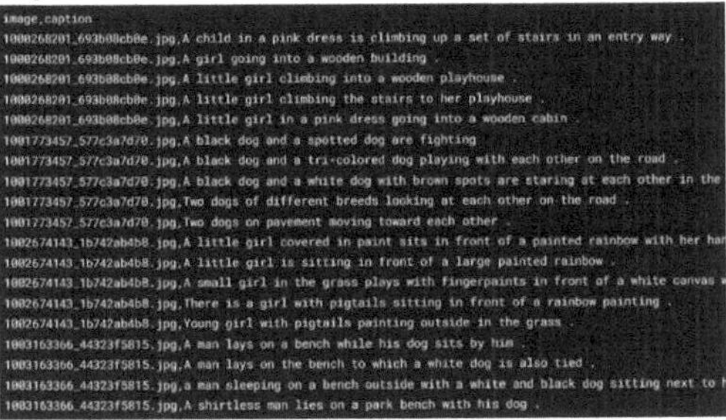

Source: (HSANKESARA, 2018) (adapted).

Description: **Figure 11** is a print screen showing the layout of the 5 captions assigned to each of the 8,000 reference images in the data set. This image was taken from the page official Flickr 8k dataset, on the kaggle platform for data science competitions.

Figure 12 - Examples of images in the Flickr 8k dataset.

Source: (HSANKESARA, 2018) (adapted).

Description: **Figure 12** refers to a print screen containing some reference images, which aim to represent different types of real-life scenarios, containing both people of different nationalities and landscapes in different countries. This image was taken from the official Flickr 8k dataset page on the kaggle platform for data science competitions.

2.7.4 Optical Character Recognition (OCR)

The emergence of computer vision was largely motivated by the need to create new and effective tools to enable people with visual impairments to enter education and the labour market in an inclusive way, which consequently led to research into this subject, more specifically into the development of specialised optical character recognition (OCR) tools. Aimed at increasing the functional capacities of people with visual impairments (Lucas at el, 2021).

Figure 13 - Representation of how OCR works.

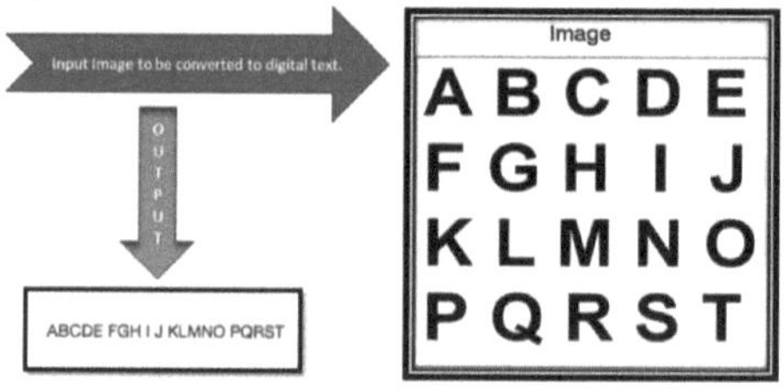

Source: (edureka, 2021) (adapted).

Description: **Figure 13** shows a representation of how OCR works, illustrating an input image containing several letters of the alphabet, then this image is digitally converted, its output showing all the letters contained in the image in digitised form.

3. DEVELOPMENT METHODOLOGY PROPOSED

The development of tools based on computer vision and natural language processing has grown increasingly, thanks to technologies built in languages such as C++ and Java, especially applications built in Python, which is responsible for abstracting a large part of the complex mathematical calculations needed for these systems to work. (mordorintelligence, 2023)

This chapter will give a practical introduction to the operation of a speech synthesiser and a speech recognition library. It will then cover the operation of the computer vision libraries OpenCV and YOLO for Pedestrian Detection and Tracking, Facial Detection and Recognition, Distance Estimation and Obstacle Anticipation, Linc Detection and Corner Detection, as well as algorithms that implement the Flickr 8k database to generate image captions. And finally, optical character recognition using the Tesseract tool.

All the codes presented in this chapter are available in the "GuiaInteligente" repository, attached to this document. It's worth noting that the database of images and videos used in this final project, contained in the "GuiaInteligente" repository, was produced as a simulation, exclusively on the premises of the Federal Institute of Education, Science and Technology of Pernambuco IFPE, Jaboatao dos Guararapes campus.

The recordings were made between 04/07/2022 and 16/11/2022 at 4 different times, all in the morning. It should be emphasised that all the people involved in the preparation of this database are visually impaired and have no significant difficulties in seeing.

In order to make it easier to understand the development methodology proposed in this work, a division has been made with regard to the order in which the tools discussed in this document are presented. This division is arranged as follows, as described in **Table 06** below.

Table 06 - Sequence of functionalities to be covered in **Chapter 03.**

OpenCV	You Only Look Once	Speech Recognition	Flickr 8k	Google Text to speech (gTTs)	Tesseract (OCR)
OpenCV	YOLO		flickr ••	TTS	Tesseract OCR

Source: Own elaboration

Description: Table 06 presents a table in order to simplify the chronological order of the contents that will be covered in the methodology proposed in this chapter.

For a good understanding of the project, it is necessary to organise the various components relating to up-to-date technologies. In this sense, the "Guialnteligente" project has been organised under the following structure of directories and files. This structure is exemplified **in Figure 14** below.

Figure 14 - Structure of the "Guialnteligente" project.

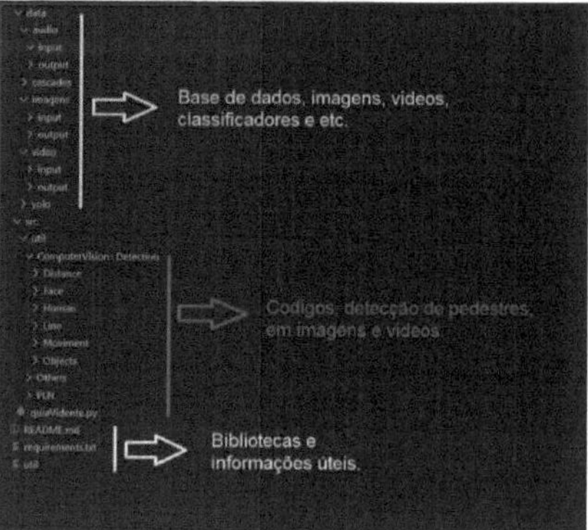

Source: Own elaboration.

Description: **Figure 14** contains a print screen of the "Guialnteligente" project, opened in the visual studio code IDE, containing the folder and subfolder structure, as well as the files used in the project.
The structure was set up to summarise the three basic elements for running the codes and obtaining their respective results using the simulated database. At the top of the image, highlighted in blue, are the images, videos and Haar Cascades classifiers. In the centre of the image, highlighted in red, are the ".py" executable codes (files in python format), and finally, at the bottom of the image, highlighted in yellow, are the project configuration information and how to install the necessary libraries.
A series of videos were recorded and, from these videos, images were generated, in different places and on different days, places that are usually accessed relatively frequently by students at the Jaboatao dos Guararapes campus, these places being the corridors, the staircases leading up to the first floor of blocks C and D.

Figure 15 - Series of videos recorded at the Jaboatao dos Guararapes campus.

Source: Own elaboration.

Description: **Figure 15** shows a print screen of a repository from the "Guialnteligente" project, designed to store the simulated database used in the tests. With regard to the practical presentation of the tools mentioned in the

Table 06, this arrangement will be presented below.

- Pedestrian Detection and Tracking.

- Face Detection and Recognition.

- Estimating Distance and Anticipating Obstacles.

- Line Detection and Corner Detection

- Flickr 8k.

- Tesseract (OCR).

It will be proposed to integrate these tools and knowledge in order to create a tool capable of offering greater autonomy and safety for visually impaired people to move around the internal premises of the Federal Institute, Jaboatao dos Guararapes campus. This system is based on the knowledge developed in **Chapter 02**, namely assistive technologies, audio description and orientation and mobility, computer vision and natural language processing, working in an integrated manner in favour of accessibility. **Figure 16** below shows the physical representation using a flowchart.

This physical representation of the tool is divided into input, processing and output devices. The input and output systems are responsible for communicating with the user and the environment, which is made possible by a processing device capable of relating the interactions by means of an electronic circuit.

Figure 16 - Physical representation of the "SmartGuide" tool.

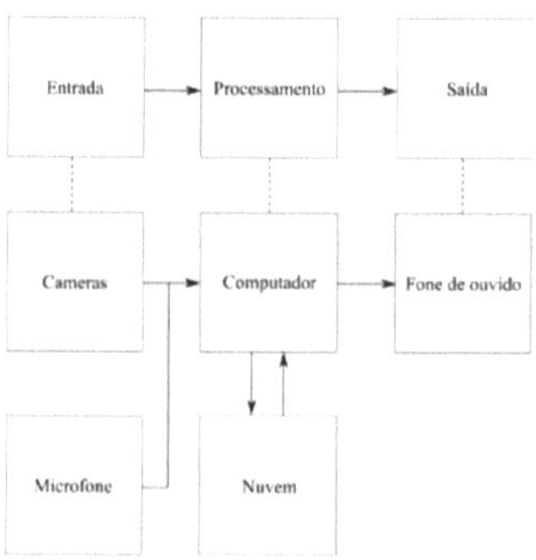

Source: (Scientia Prima, 2020) (Adapted).

The logical representation of the tool, described in **Figure 16**, is responsible for interpreting the visual signals/recordings coming from the environment to be recognised, converting this data into audio signals by means of a voice synthesiser, which is responsible for interacting with the user. It interprets the sound information obtained and make the best description of the journey based on this information, taking into account, above all, the safety and physical integrity of the user.

Figure 17 - Logical representation of the "SmartGuide" tool.

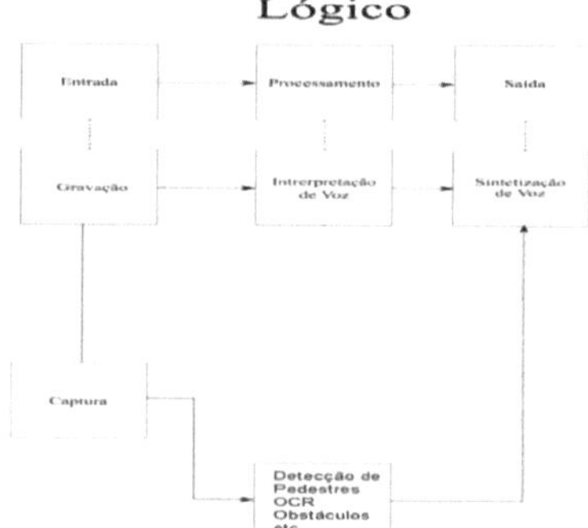

Source: (Scientia Prima, 2020) (Adapted).

Description: **Figure 17** contains a flowchart representation of the logical structure of the "GuiaInteligente" project.

3.1 Design Science Research (DSR)

Design Science is the paradigm for a research method that supports the development of innovative artefacts for solving overly complex problems in the field of information systems (IS). Most of the theoretical work based on DSR for scientific research is contained in the field of information systems.

"Area of knowledge that deals with the planning, programming and design of the objects with which man deals in his daily life, as well as the environments in which he maintains his living space." (Escola Superior de Design Industrial da UERJ - ESDI)

According to Alan Hevner et al, 2004, DSR research is based on 7 fundamental guidelines:

- Guideline 1: Design as an artefact.

- Guideline 2: Relevance of the problem.

- Guideline 3: Evaluation in Design Science Research.

- Guideline 4: Research contribution.

- Guideline 5: Rigour in Design Science Research.

- Guideline 6: Design as a search process.

- Guideline 7: Communicating results.

Following the guidelines stipulated by Alan Hevner et al, 2004, it is necessary to carry out a primary survey of the materials and methods that must be employed for the proper development of the project, created on the basis of the knowledge acquired during the creation of this course completion work, in **Table 07** below, these elements will be described.

Table 07 - Resources used or for future integration.

Items	Used
Notebook samsung essentials e30 (poss)	Yes
Bone conduction headset (poss)	Yes
Smartphone LG k22 (poss)	Yes
Webcam (available)	Yes
Raspberry pi (None)	No
glasses with integrated bone conduction headset and attached micro camera (not available)	No

Source: Own elaboration

Description: **Table 07** lists a series of items used in the development of the "GuiaInteligente" project.

3.2 Synthesis and speech recognition

In this section, the main points regarding the operation of the gTTs (Google Text-to-Speech) library for text conversion will be demonstrated. in speech TTs (Text-to-Speech), the speech_recognition library will then be used to recognise speech. In this sense, from the detections resulting from computer vision methods, which will be discussed in the following sections, texts with pedestrians, faces, distance, obstacles, lines and corners were detected, as well as texts using optical character recognition.

To use the gTTs library, you first need to install the library using the pip command "pip install gTTs", then import the library as follows "from gtts import gTTs", in the way the project was built, from the text assigned for testing, the code saves the file in ".mp3" format and runs it simultaneously. The voice synthesis is Google's own and in Portuguese format.

Figure 18 - Code implementing the gTTs library.

```python
# Import the required module for text
# to speech conversion
from gtts import gTTS

# This module is imported so that we can
# play the converted audio
import os

# The text that you want to convert to audio
mytext = 'Guia Inteligente'

# Language in which you want to convert
language = 'pt-br'

myobj = gTTS(text=mytext, lang=language, slow=False)

# Saving the converted audio in a mp3 file named
myobj.save("data\\audio\\output\\audiodescricao.mp3")

# Playing the converted file
os.system("data\\audio\\output\\audiodescricao.mp3")
```

Source: (Geeksforgeeks, 2023) (Adapted).

Description: **Figure 18** shows a print screen of the "text_to_voice.py" code. The variable that receives the text for which voice synthesis is desired contains the following phrase: "Intelligent Guide", and the variable that receives the desired language for the synthesiser result contains the Multilingual Codex: "en-br".

3.3 Detection and Tracking of Pedestrians

In this section, the main points required for **pedestrian detection and tracking** using the Histograms of Oriented Gradients (HOG) method will be demonstrated, followed by the implementation of the Haar Cascades classifiers. **Figure 19** below shows the sequence of codes needed to implement the HOG method.

Figure 19 - Code implementing the HOG method in the Python programming language.

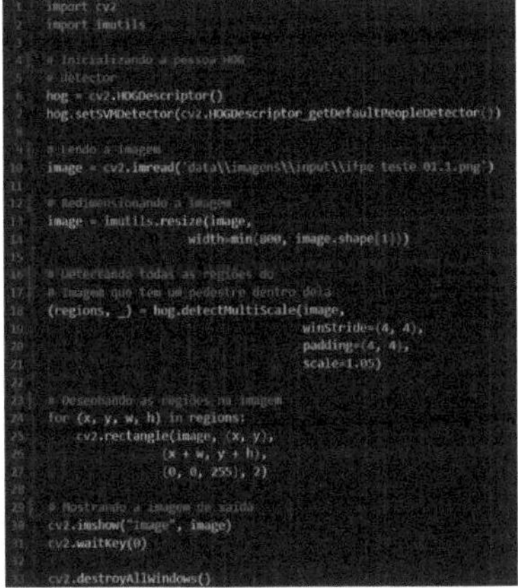

Source: (Geeksforgeeks, 2023) (Adapted).

Description: Figure 19 contains a print screen of the code "image_pedestrian_detection_01.py", which implements the set of HOG methods for detecting and tracking pedestrians.

Figure 20 - Multiple Bounding boxes generated in the images.

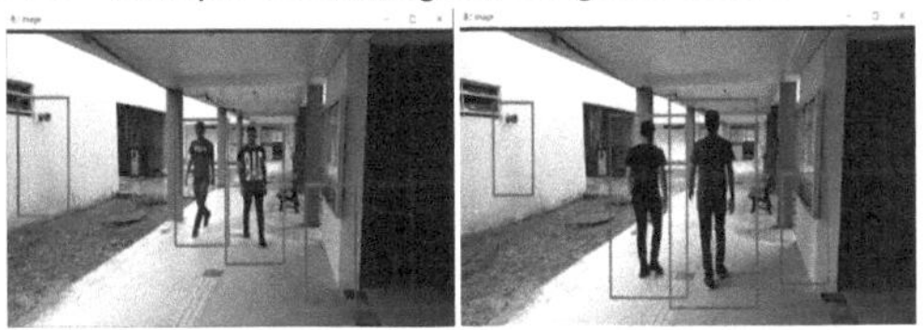

Source: Own elaboration.

Description: Figure 20 shows a print screen using the "cv2.imshow" method, where two pedestrians/students were detected walking together in the access corridor to block D of the Federal Institute, Jaboatao dos Guararapes campus. In the first image, the students are facing the camera, while in the second image, the two students are facing away from the camera.

Evaluating the results for this particular detection, we realise that the occurrence of bounding boxes in false positives is probably due to factors external to the HOG method. The factors for this result are diverse: the low light levels in the environment, given the time at which it was recorded, the low quality of the camera used to capture these images, are examples of factors that can significantly influence the results obtained.

With regard to the procedures required for Pedestrian Detection and Tracking using Haar Cascade classifiers. **Figure 21** below shows the different types of classifiers available. Each Haar Cascade classifier corresponds to a different and specific type of detection, which means, for example, that the "haarcascade_eye.xml" file was trained specifically to detect eyes, so its use to detect other elements of the human body was unsuccessful.

Figure 21 - Different types of Haar Cascades classifiers available.

Source: Own elaboration.

Description: **Figure 21** contains a print screen of the "GuiaInteligente" project, with the Haar Cascades classifiers imported into the project.

Figure 22 - Code that implements the Haar Cascades Classifiers in the language Python programming.

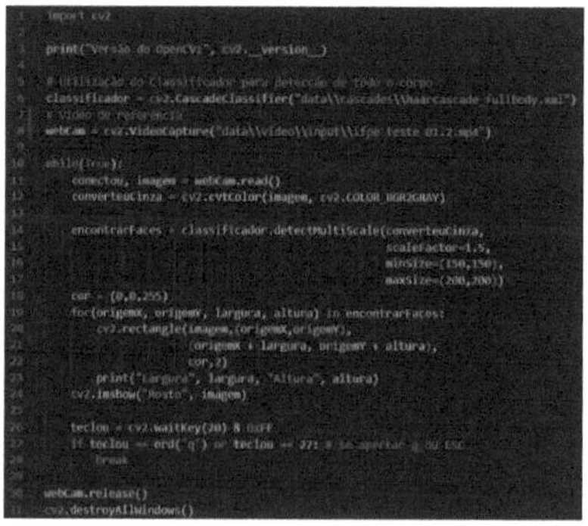

Source: Own elaboration.

Description: **Figure 22** contains a print screen of the code "image_pedestrian_detection_02.py", which implements the Haar Cascades classifiers for pedestrian detection and tracking.

Figure 23 - Bounding box containing the detection of two students.

Source: Own elaboration.

Description: **Figure 23** shows a print screen of the "cv2.imshow" method, detecting two pedestrians/students. Two bounding boxes were generated for each student detected in the video. These students are walking down the stairs on the first floor of block D on the Jaboatao dos Guararapes campus. From the perspective from which the image was generated, the students are facing the camera.

One factor that may have had a significant impact on the results obtained is the limited number of images used to generate the classifier, given that the ".xml" files used in the tests are part of a database of pre-prepared models available free of charge on the internet.

With regard to how YOLO works for detecting and tracking pedestrians, it's important to note that, due to the tool's domain issues, the tests with the library in question were carried out exclusively using statistical images from videos produced at the Federal Institute, Jaboatao dos Guararapes campus.

The sequence in which **Figures 24**, **25** and **26 are** laid out refers respectively to a test image generated from a video, a print screen of the code implemented on the Google Colab platform and finally, a print screen of the Google Colab tool console with the results.

Figure 24 - Image uploaded to the Google Drive platform.

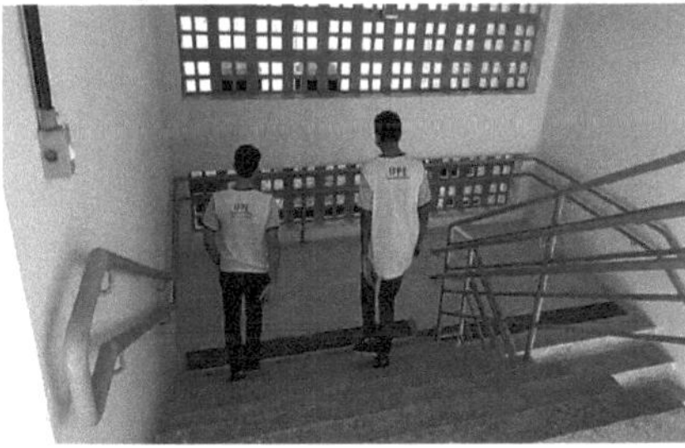

Source: Own elaboration.

Description: Figure 24 shows two students/pedestrians walking down the stairs of the first floor of block C of the Federal Institute, Jaboatao dos Guararapes campus. From the perspective from which the image was generated, the students are facing away from the camera.

Figure 25 - Code implementing YOLO on the Google Colab platform.

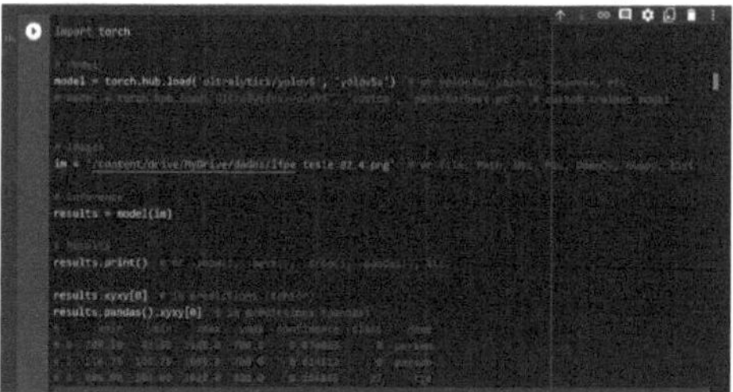

Source: Own elaboration.

Description: Figure 25 shows a print screen of the code, implemented in the Google Colab tool.

Figure 26 shows the results displayed on the Google Colab console.

Figure 26 - Result for YOLO detection on the Google Colab platform.

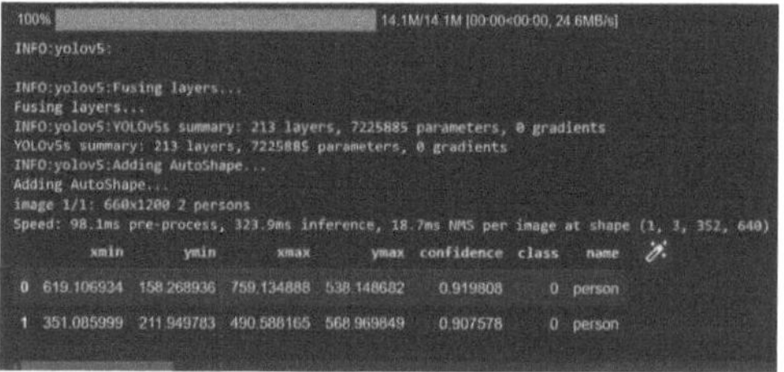

Source: Own elaboration.

Description: **Figure 26** shows a print screen with the result of running the code shown in **Figure 25** and detecting two pedestrians/students.
Figure 26 shows that two people were found in the image and that the degree of confidence in percentage for these detections is 91 per cent for vector [0] and 90 per cent for vector [1].

3.4 Face Detection and Recognition

This section will demonstrate the operation of the OpenCV computer vision library and the Haar Cascades classifiers implemented in the Python programming language for the detection and facial recognition of students, staff and visitors to the Federal Institute, Jaboatao dos Guararares campus.

Figure 27 - Detecting a student's face by generating a bounding box.

Source: Own elaboration.

Description: **Figure 27** shows a print screen of the "cv2.imshow" method. In the image in question, the student/pedestrian is walking down the corridor to the access ramp to the first floor of block D of the Federal Institute, Jaboatao dos Guararapes campus.

3.5 Estimating Distance and Anticipating Obstacles

This section will demonstrate the operation of the OpenCV computer vision library and the Haar Cascades classifiers, implemented in the Python programming language, in order to estimate the distance between a predefined reference object and a target object. In the interests of simplification, and in order to generate a greater number of successful cases for distance estimation, the "haarcascade_frontalface_default.xml" classifier was used, focusing on detecting a person's frontal face.
The choice to use this particular classifier arises from the fact that, as a disabled user, using the device (micro camera) at eye level will make it easier to detect the faces of other people who are within the capture device's field of vision.

Figure 28 - Test image used to estimate distance.

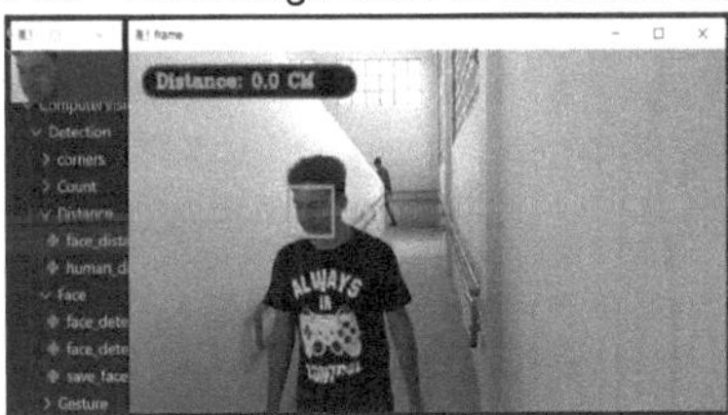

Source: Own elaboration.

Description: **Figure 28** shows a print screen of the "cv2.imshow" method, where two images were generated, one containing a static image, with the face used as a reference, and the other, relating to a specific moment in the video, where the reference image used was found, indicating the distance from the camera to the object in question, which in this case is a face.

In order to anticipate possible obstacles when travelling, OpenCV's native gaussian blur method was used, as well as the numpy library for processing matrices and mathematical functions, to structure the code, and the result, with the test database, produced on the premises of the Federal Institute, Jaboatao dos Guararapes campus, can be seen in **Figure 29** below.

Figure 29 - Test image for anticipating obstacles.

Source: Own elaboration.

Description: **Figure 29** shows a print screen using the "cv2.imshow" method. Due to the use of the Gaussian blur method, the image is displaced, in the upper part of the image there is text written: "On the left", and the student/pedestrian in the image is in fact in this position.

3.6 Line Detection and the Detection of Corners

This section will demonstrate the operation of the OpenCV computer vision library and the Hough transform technique, implemented in the Python programming language. Simply put, Hough transform is a mathematical technique specialised in detecting geometric shapes in digital images.

Figure 30 - Test image used for line detection.

Source: Own elaboration.

Description: **Figure 30** shows a print screen of the "cv2.imread" method, where several old lines were generated, where the algorithm considered the incidence of predefined lines in the image in question.

The image shown in **Figure 30** is the result of applying a method known as "probabilistic line transformation", which was applied to the same image shown in **Figure 01**, inserted at the beginning of this document.

In order to detect regions in images with a large variation in intensity in all directions, Chris Harris and Mike Stephens developed a method for detecting corners in their article A Combined Corner and Edge Detector in 1988, which is now known as the Harris Corner Detector.

Figure 31 - Practical application of the Harris Corner Detector.

Source: Own elaboration.

Description: **Figure 31** shows a print screen of the "cv2.imshow" method, where the image of the façade of the library of the Federal Institute, Jaboatao dos Guararapes campus, generated several red dots where the Harris Corner method identified large variations in intensity.

3.7 Flickr 8k

This section will demonstrate how the Flickr 8k dataset works to generate image captions. The results presented in this section, are the result of the work carried out by Raman Shinde in his article entitled, "Image Captioning With Flickr 8k Dataset & BLEU", available in full on Medium, the link to which is attached to this document. The elements covered in the aforementioned article summarise some elementary steps for using the database correctly.

• Image characterisation.

• Pre-processing of subtitles.

Characterisation uses a pre-trained model from the Imagenet dataset provided by keras, containing 14 million images in the dataset, and more than 21,000 distinct class groups. In the aforementioned work, InceptionV3 provided by Google was used, providing a smaller file size of 96 MB and being faster to train. It is worth noting that it is necessary to resize all the images to the size (299 x299).

During pre-processing, the captions are read directly from the "Flickr8k.token.txt" file and the captions are added to the k:v dictionary, where k = image identifier (id) and v = value (caption list). The decoding for the dictionary takes place as follows: "startseq " + caption + " endseq".

• startseq : This will act as our first word when the image vector extracted from the resource is fed to the decoder. This will start the subtitle generation process.

• enseq : This tells the decoder when to stop. We will stop predicting words as soon as endseq appears or we have predicted all the words in the train dictionary, whichever comes first.

Figure 32 - Pre-processing the captions in Flickr 8k.

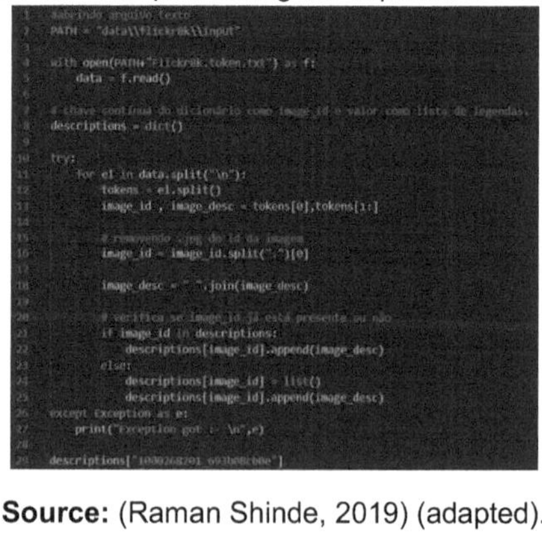

Source: (Raman Shinde, 2019) (adapted).

Description: **Figure 32** shows a print screen of the code used to pre-process the captions assigned to each of the 8,000 images in the Flickr 8k dataset.

Figure 33 - Visualisation of the captions assigned to the reference images.

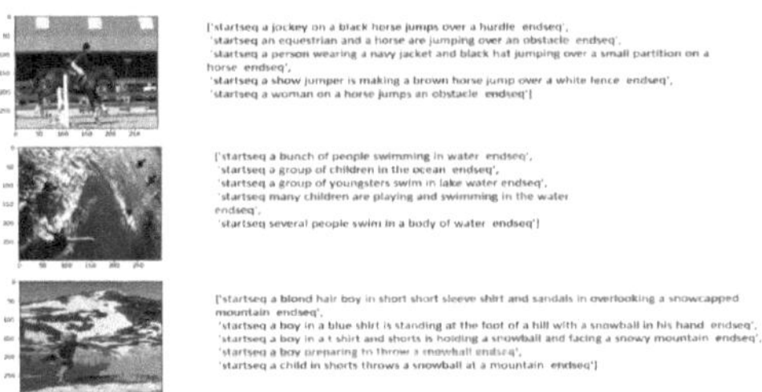

Source: Raman Shinde (adapted).

Description: **Figure 33** shows a print screen of the execution screen of the notebook created in Google Colab, containing three images from the reference set and five captions associated with each of these images.

3.8 Tesseract (OCR)

This section will demonstrate how Tesseract works for optical character recognition, implemented in the Python programming language. In Python specifically, Tesseract is used through the PyTesseract library. In the example shown in **Figure 34** below, the method used was "pytesseract.image_to_data" to identify regions of interest, and the "cv2.rectangle" method was used to generate the bounding boxes, together with a sorting algorithm.

Figure 34 - Multiple bounding boxes generated in the reference image.

Source: Own elaboration.

Description: **Figure 34** shows a print screen of the "cv2.imshow" method. In the image in question, multiple bounding boxes were generated, with the occurrence of true positives and false positives.

This result, shown in **Figure 34**, was obtained by executing the file "tesseract_02.py", where the method "pytesseract.image_to_string" was also used, resulting in the detection of scattered parts of words. Words such as: "ARIANO SUASSUN", "INSTITUTO FEDERAL", "Campus Jaboatao dos", among other single letters.

4. ANALYSING THE RESULTS

This chapter will present the results of the evaluation carried out through the proposal to build an accessibility tool using the natural language processing libraries, Speech_recognition, as well as the computer vision libraries OpenCV, YOLO for pedestrian detection and tracking, among other important topics, in order to build an accessibility tool for visually impaired people, observing the fundamentals of orientation and mobility, simultaneous audio description using a voice synthesiser.

The operation of the Flickr 8k database for the automatic generation of image captions will also be discussed, and it should be emphasised that the results presented, exclusively in relation to this tool, are results referring to the execution of the work developed by Raman Shinde in his article entitled, "Image Captioning With Flickr 8k Dataset & BLEU". (Raman Shinde, 2019)

4.1 Speech_recognition and Google Text-to-Speech (gTTS)

This section presents the results of tests carried out using the natural language processing library gtts to convert digital texts into audio. During the tests with the aforementioned library, an excellent result was found for synthesising the human voice, and the choice was made to generate the data in the Portuguese language 'pt-br'. A series of phrases were drawn up, applied to the context of this course conclusion, which is the internal premises of the Federal Institute, Jaboatao dos Guararapes campus. Examples of phases such as:

- "One person is 6 metres away."

- "Two students are just ahead."

- "Continue on the tactile floor."

- "Obstacle on the right."

- "Obstacle on the left."

These sentences are in the context of what is intended as the end result of using these two libraries, in a context in which all the knowledge covered so far works in a 100 per cent integrated way, namely the Speech_recognition library and Google Text-to-Speech (gTTs).

Specifically, in the tests with the Speech_recognition library, the notebook mentioned in the materials and methods of this final project, and detailed in chapter 03 section 3.1, was used. In summary, it was possible to observe that the use of the tool resulted in excellent results, something close to 100% (one hundred per cent) assertiveness for the detection of spoken words/test phrases.

With regard to the Google Text-to-Speech (gTTs) library for synthesising your voice, the metrics and results remained practically the same, with the voice generated by the synthesiser being almost 100% (one hundred percent) audible, clear and precise.

4.2 OpenCV

The following subsection will present the results of the tests carried out with the OpenCV computer vision library using HOG (Histograms of Oriented Gradients) methods, as well as the results obtained using Haar Cascades classifiers, among other methods common to the proper functioning of these two approaches.

4.2.1 Histogram of Oriented Gradients (HOG)

The results presented below refer to the execution of the "video_depestrian_detection_01.py" code, present in the "GuiaInteligente" project. **Figure 35** shows a graph with the results for the detections carried out with the simulated database, on the internal premises of the Federal Institute, Jaboatao dos Guararapes campus.

Figure 35 - Graphical representation of the results with the HOG method.

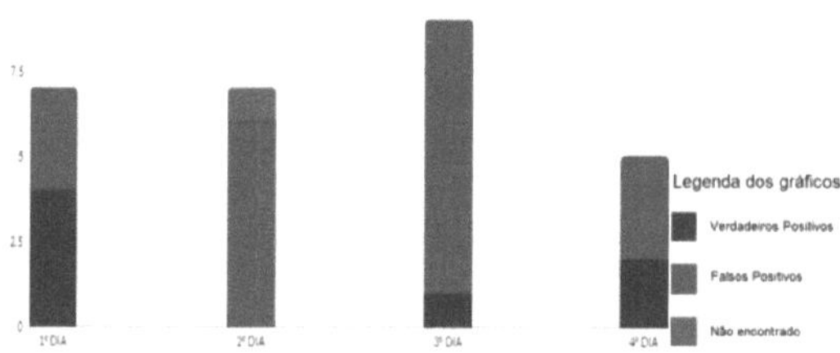

Source: Own elaboration.

Description: Figure 35 shows the sum of the incidence of true positives and false positives for the detections, as well as the cases in which no pedestrians were found, over a four-day period.

The general aim of the tests carried out was to assess how significantly the elements of the environment can interfere with the assertiveness of pedestrian detection and tracking. Issues such as brightness (sunlight or artificial light), shadows in corridors and stairwells, as well as the quality of the equipment/camera used for the recordings, all affect the final result.

4.2.2 Haar Cascades

The results presented below refer to the execution of the "video_depestrian_detection_02.py" code present in the "GuiaInteligente" project. **Figure 36** shows a graph with the results of these detections, carried out on the internal premises of the Federal Institute, Jaboatao dos Guararapes campus.

Figure 36 - Graphical representation of the results with Haar Cascades.

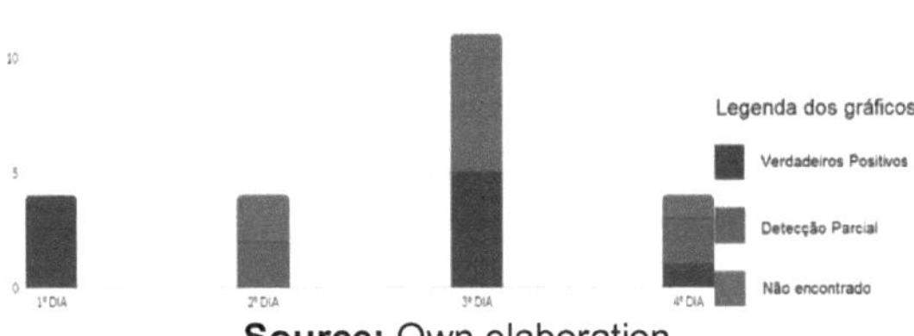

Source: Own elaboration.

Description: Figure 36 shows the sum of the incidence of true positives and false positives for the detections, as well as the cases in which no pedestrians were found, over a four-day period.

The general aim of these tests was to assess the incidence of positive and negative detection cases, as well as the occurrence of cases in which no pedestrian/student could be detected.

Comparing the two graphs shown in **Figures 35 and 36**, it can be seen that in the graph shown in **Figure 36**, there are more cases of non-detections, which at first may indicate that the Haar Cascade classifier is inferior to the HOG method, which does not necessarily correspond to reality, given that, as mentioned above, elements of the environment can have a major influence on the final result.

4.3 You Only Look Once (YOLO)

The results presented below refer to the execution of the adapted code on the Google Colab platform with the aim of implementing the use of the YOLO neural network model for pedestrian/student detection and tracking. **Figure 37** shows a graph with the results for these detections, carried out on the internal premises of the Federal Institute, Jaboatao dos Guararapes campus.

Figure 37 - Graphical representation of the results with YOLO.

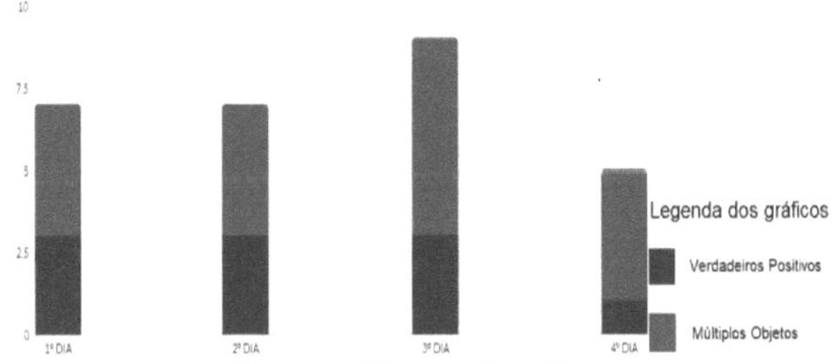

Source: Own elaboration.

Description: **Figure 37** shows a graph representing the sum of the true positives (in blue) for pedestrian detections specifically, where green highlights the cases in which multiple objects (not pedestrians) were detected in the images.

The general objective of the tests carried out using YOLO was to evaluate, in an introductory manner, the multiple detections in the images produced in the internal premises of the Federal Institute, Jaboatao dos Guararapes campus.

4.4 Flickr 8k

The results presented below refer to the execution of the work developed by Raman Shinde in his article entitled, "Image Captioning With Flickr 8k Dataset & BLEU", this same project was covered in the previous chapter, therefore, in this section, the results of the tests carried out in this document will be analysed. (Raman Shinde, 2019)

Figure 38 - Example output for generating subtitles.

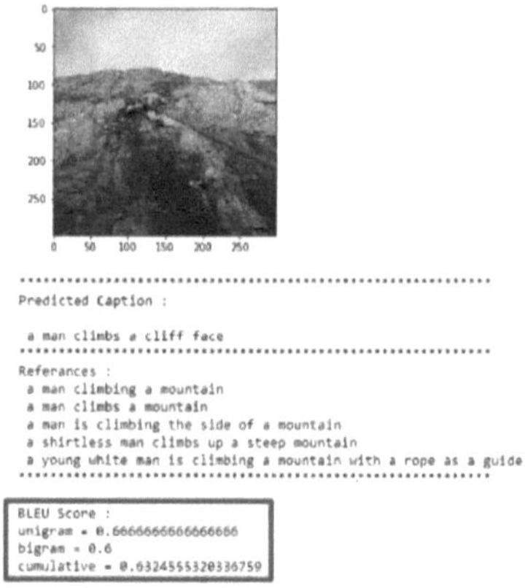

Source: (Raman Shinde, 2019) (Adapted).

Description: **Figure 38** shows a print screen, an example of a safda for generating image captions. In general, the performance of the model used to obtain the results could be improved by training a larger data set and adjusting the hyperparameters more precisely. Above, you can see, highlighted in red, that the BLEU unigram scores favour short forecasts.

4.5 Tesseract (OCR)

The results presented below refer to the execution of the code, "tesseract_02.py", which evaluated the performance of the library for converting digital images into digital texts, known as optical character recognition (OCR), applied to the internal premises of the Federal Institute, Jaboatao dos Guararapes campus. In order to clearly visualise the integration of the various skills developed in this course, the tests for optical character recognition were carried out using the Tesseract library. in line with the gTTS library, for recognising texts and synthesising them into audio, simultaneously, in real time. As such, speech synthesis and comprehension is dependent on the quality of the reading of the characters/words detected in the images. As a result, the percentage for these detections is uncertain because, in the simulated test database, no signs or banners with text were registered at the Federal Institute, Jaboatao dos Guararapes campus, only the façade of the campus library.

5. FINAL CONSIDERATIONS AND FUTURE WORK

In this final chapter, final considerations will be presented on the use of the natural language processing libraries Speech_recognition and Google Text-to-Speech (gTTs), as well as the use of the computer vision library OpenCV, specifically, using the HOG (Histograms of Oriented Gradients) method and Haar Cascades classifiers, applied to the test database, produced as a simulation on the premises of the Federal Institute, Jaboatao dos Guararapes campus. The operation of the YOLO computer vision tool will also be discussed. The operation of Flickr 8k, used to generate image captions, based on the results presented by Raman Shinde in his article entitled, "Image Captioning With Flickr 8k Dataset & BLEU". And finally, the operation of the Tesseract library for optical character recognition (OCR). (Raman Shinde, 2019)

5.1 Final considerations

In this work, the main concepts relating to visual impairment were discussed, as well as the challenges that visual impairments impose on people's lives. The guiding concepts relating to orientation and mobility (OM) were also addressed. This was followed by a discussion of assistive technologies in order to increase the functional capacities of these people, and then a discussion of audio description of scenes. Two themes common to natural language processing were discussed further on, and finally a more in-depth discussion of the guiding elements of natural language processing. Based on these concepts, a development methodology was proposed which had as its premise the complete integration of these resources in order to promote accessibility for visually impaired people by developing a specific tool for this purpose. It is important to emphasise that, due to financial constraints, all of the work carried out in this project was not carried out in the context of the project. The codes developed so far are fully and publicly available on the github platform attached to this document, making it possible to measure the feasibility of building an accessibility tool for visually impaired people.

5.2 Limitations

The proposal to develop an accessibility tool for people with visual impairments, discussed in this final project, is limited in some respects to financial and budgetary issues that are too costly for the author. Another aspect that should be noted concerns the metrics used to assess the degree of assertiveness of the algorithms, methods and tools used during the tests with the simulated database, on the premises of the Federal Institute, Jaboatao dos Guararapes campus. In this regard, it is important to note that the recordings took place exclusively in the morning, given the compatibility of schedules between the author of this document and the students duly cited in the acknowledgements of this document.

5.3 Future work

Some important points could be developed in future proposals, based on the topics covered in this document, thus making it possible to improve pedestrian detection and tracking using the computer vision libraries OpenCV and YOLO, as well as the generation of image captions using the Flickr 8k database. Specifically with regard to OpenCV, the Haar Cascades classifiers, a larger number of classifiers can be explored.To study ways of counting objects/people in an image, as well as the partial recognition of human gestures and the recognition of human features. To study the advances that robotics combined with computer vision could bring to the topics covered in this document, such as, the use of geolocation and the delimitation of a route, integrated with tools such as Google Maps.Another point that could be considered in future work is the use of elements addressed in other proposals for accessibility tools for people with visual impairments, proposals that show results in different scenarios and situations from those addressed in this final project. Due to the large data set, the complete integration of all the technologies mentioned in this document could end up resulting in a scenario where a huge amount of information, such as obstacles ahead, distance between the user of the device and a pedestrian/student, information signs, tactile surfaces, among other elements, is detected simultaneously, among other elements, detected simultaneously, it should be noted, the creation of elements based on machine learning and deep learning, in order to ABSTRACT this huge amount of information, in order to create a "logical and situational line", in order to guide the disabled person in a safe and determined way.

Another important point concerns the simulated database used in the tests, which should generate new images/videos of the campus rooms and laboratories, as well as the detection of posters, signs and banners displayed in the Federal Institute.

REFERENCES

ALVES, MARIA EDUARDA SOFFIATI, LIMA, FERNANDO PAIM. **Detection and facial recognition from a database**, IFMG - Campus Formiga - 2019.

Alexssandro Ferreira Cordeiro, Cristian A. U. Ojeda, Pedro L. P. Filho, Gustavo R. Valiati, **Real-time pedestrian tracking and counting using digital images**, LATINOWARE 2019, XVI Latin American Congress on Free Software and Open Technologies, 2019.

ALMEIDA, M.G.S. **Instituto Benjamin Constant: 160 years of inclusion**. Benjamin Constant Magazine, 2014.

BERNARDES, **Junior et al. System helps visually impaired people get around**. 2016.

BRAZIL. **Constitution** (1988). **Constitution of** the Federative Republic of Brazil. Brasflia, DF: Senado **Federal**: Centro Grafico, 1988.

Bradski, Gary; Kaehler, Adrian. **Learning OpenCV: Computer Vision with the OpenCV Library**, 2008.

BITTENCOURT, Z. Z.; HOEHNE, E. L. **Quality of life of visually impaired people**. Medicina (Ribeirao Preto. Online), [S.l.], v39, n.2, p.260, 2006.

Boland, R. Design in the punctuation of management action. Managing as designing: **Creating a vocabulary for management education and research** (2002), 106-112.

Dr Vipulsangram.K.Kadam, Deepali G. Ganakwar, Face Detection: A Literature Review, Maharashtra, India, IJIRSET, July 2017.

D. Jay Gense, En.S., Marilyn Gense, M.A., **The Importance of Orientation and Mobility Skills for Students who are Deaf-Blind**, Helen Keller National Centre Perkins School for the Blind Teaching Research, 2004.

ELIAS AUGUSTO FANK, DENIO DUARTE. **Using computer vision and character detection to help visually impaired people navigate indoors**, UFFS - Campus Chapecô, 2016.

Flavio Augusto Schiave Germano et al, **Study of the causes of blindness and low vision in a school for the visually impaired in the city of bauru**, 2019

FELIPPE, J.A. de M.; FELIPPE, V. L. R.. **Orientation and mobility**. Sao Paulo: Laramara - Associação Brasileira de Assistência ao Deficiencia Visual, 1997

FELIPPE, J.A. de M. Caminhando juntos: manual of basic orientation and mobility skills. Sao Paulo: **Laramara**, 2001.

FRANCO, Eliana Paes Cardoso; SILVA, Manoela Cris na Correia Carvalho. Audiodescription: A brief historical overview. In: MOTTA, Lfvia Maria Villela de Melo; ROMEU FILHO, Paulo (Orgs.). **Audio description: transforming images into words**. Sao Paulo: Secretariat for the Rights of People with Disabilities of the State of Sao Paulo, 2010.

GOMES, D. dos S. **Artificial Intelligence: concepts and applications**. Olhar Cientffico. v1, n. 2, p. 234-246. 2010.

GABRIELA P, MARIANTO, JOYCE A. P. SOARES, EDUARDO M. A. AMARAL, **Victim detection and rescue system for an autonomous line-following robot**

based on computer vision. National Robotics Exhibition (MNR), Federal Institute, Serra campus.

Hyun-Tae Kim, Sang-Hyun Lee, **A study on object distance measurement using OpenCV-based Yolo 5**, Computer Engineering, Homan University, Korea, 2021.

Ing-Jr Ding, Chih-Ta Yen, and Yen-Ming Hsu, **Developments of Machine Learning Schemes for Dynamic Time-Wrapping-Based Speech Recognition**, Hindawi Research Article, National Formosa University - Huwei, Taiwan, 2013.

Joao Alvaro de Moraes Felipe. **Caminhando juntos: manual das orientações bâsicas de orientaçao e mobilidade**. governo do estado de sao paulo. secretaria de estado dos direitos da pessoa com deficiência. sao paulo, 2009. Available at: <http://visaosubnormal.org.br/downloads/serie_deficiencia_visual_vol4_cbo_bq.pdf>

Souza C.A. (2003) **"Development of a Computer Vision System for Servo-Visual Control Applications of Mobile Robots"**. Monograph. SBC University, Salvador.

MARQUES FILHO, O.; VIEIRA **NETO**, H. **Digital image processing**. Rio de Janeiro: Brasport, 1999.

SILVA, J. B. **Exploring the Viola-Jones algorithm in facial detection and recognition**. 2018. 91 f. Coursework (Computer Engineering) - Federal University of Sao Carlos, Sao Carlos, 2018.

Scientia Prima, v. 6, n. 1, p. 98-116, **System to help visually impaired people get around**. May 2020.

MOTA, Maria Glôria B. **Orientation and mobility: basic knowledge for the inclusion of visually impaired people**. Brasília: **MEC**, 2003.

WILLIAM MARRION CAVENAGLI, CAROLINE MAZETTO MENDES, **Detection of parking spaces using computer vision and convolutional neural networks**. Positivo University, 2020.

Lucas K. Uezima, Henrique Y. Y. Nishimoto, Rafael Barbosa, Prof. Dr Antonio Luiz Basile, **Using Tesseract OCR for handwritten text recognition**, Mackenzie Presbyterian University (UPM) - Sao Paulo, SP, Faculty of Computing and Informatics, 2021

Intelligent robotics: **Using computer vision to control and navigate autonomous mobile robots**. Institute of Mathematical and Computer Sciences (ICMC), USP/Sao Carlos - MNR National Robotics Exhibition, 2012.

Milena L. dos Santos, Claudio R. M. Mauricio, Valéria N. dos Santos, Fabiana F. F. Peres**, Convolutional neural network for detecting pedestrians crossing at risk**, State University of Western Paraná - Centre for Engineering and Exact Sciences, Foz do Iguaçu.

Paul Viola, Michael Jones, **Rapid Object Detection using a Boosted Cascade of Simple Features**, Accepted Conference on Computer Vision and Pattern Recognition, 2001.

Li Yi-bo, LI Jun-jun a, **Harris Corner Detection Algorithm Based on Improved Contourlet Transform**, Shenyang University of aeronautics and Astronautics- China, 2011.

Renato Martins Redovalio Ferreira, Ruth Maria Mariani Braz, **Orientation and Mobility as a Practice for the Inclusion of Visually Impaired People**, , Conedu VII

National Congress of Education, October 2020.

Loiane Maria Zengo, et al, **Reports of Blind Students on the Use of the Visible Guide as a Locomotion Strategy in School Environments**, Faculty of Philosophy and Sciences - UNESP, Marflia Campus, Faculty of Higher Education of the Interior Paulista – FIAP, 2017.

Joseph Redmon, Ali Farhadi, et al, **You Only Look Once: Unified, Real-Time Object Detection**, University of Washington, 12 Nov 2015

Mai Thanh Nhat Truong, Sanghoon Kim, **A Review on Image Feature Detection and Description, Hankyong National University**, 2016

Kai Zhao et al, **Deep Hough Transform for Semantic Line Detection**, IEEE Transactions on Pattern Analysis and Machines Intelligence, May 2021.

Navneet Dalal, Bill Triggs, HAK archives-ouvertes, **Histograms of Oriented Gradients for Human Detection**, Dec 2010

Itunu A. Bamidele Ph.D, **Information Needs of Blind and Visually Impaired People**, Babcock University, Ilisan - Remo, gun State, Nigeria

MOTTA, Lfvia Maria Vilella de Mello. **AUDIO DESCRIPTION AT SCHOOL: OPENING UP WAYS OF READING THE WORLD**. 2016. Available at: <http://www.vercompalavras.com.br/pdf/a-audiodescricao-na-escola.pdf>

MOTTA, Lfvia Maria Vilella de Mello; ROMEU FILHO, Paulo (org). **Audiodescription transforming images into Words**. Sao Paulo: Secretariat for the Rights of People with Disabilities of the State of Sao Paulo. 2010.

Nilton Pinto Ribeiro Filho, **COMPUTATIONAL VISION: A NEW FIELD OF RESEARCH IN VISUAL COGNITION**, 2022. Available at: <https://periodicos.unb.br/index.php/revistaptp/article/download/17018/15504/28764>

Raman Shinde, **Image Captioning With Flickr 8k Dataset & BLEU**, 2019. Available at:<https://medium.com/@raman.shinde15/image-captioning-with-flickr8k-dataset-bleu- 4bcba0b52926>

Institute of the Blind, visual impairment, 2023. Available at: <https://www.institutodoscegos.org.br/deficiencia-visual>

WHO, First global report on blindness, WHO says world could prevent half the cases, 2019, Available at: <https://news.un.org/pt/story/2019/10/1690122>

ICF, International Classification of Functioning, Disability and Health, 2001. Available at: <https://www.periciamedicadf.com.br/cif2/cif_portugues.pdf>

Audiodescription Blog, Audiodescription, 2023. Available at: <https://www.blogdaaudiodescricao.com.br/audiodescricao>

eveo, Learn all about artificial intelligence, 2023, Available at: <https://blog.eveo.com.br/tudo-sobre-inteligencia-artificial>

Calebe Augusto dos Santos, Gino Dornelles, Text-To-Speech: Voice Synthesiser for Documents as a Support Tool for the Visually Impaired, 2008, Available at: <https://repositorio.ufsc.br/bitstream/handle/123456789/184450/TCC_Calebe_Gino.pdf?sequence=-1&isAllowed=y>

IBGE, SNIG - National Gender Information Survey, 2012. <https://cidades.ibge.gov.br/brasil/pe/recife/pesquisa/11/0>

WHO, World Report on Disability 2011, 2021. Available at:

<https://www.who.int/teams/noncommunicable-diseases/sensory-functions-disability-and-rehabilitation/world-report-on-disability>

Saude Plena, Brazilian Council of Ophthalmology warns of avoidable cases of blindness, 2019, Available at:
<https://www.uai.com.br/app/noticia/saude/2019/07/06/noticias-saude,248471/consel ho-brasileiro-de-oftalmologia-alerta-para-casos-de-cegueira-evit.shtml>

Assistiva Tecnologia e Educaçao, What is Assistive Technology, 2023, Available at:
<https://www.assistiva.com.br/tassistiva.html#:~:text=Tecnologia%20Assistiva%20%C3%A9%20the%20term,promover%20Vida%20Independente%20e%20Inclus%C3%A3o>

ADA, Americans with Disabilities Act, 2023. Available at: <https://www.ada.gov/>

Colégio gtm, Braille language: learn more about the creation of this reading system, 2023. Available at:
<https://www.colegiotgm.com.br/2021/04/08/linguagem-braille-criacao/>

Enap, National School of Public Administration, 2023. Available at:
<https://enap.gov.br/pt/>

Jéssica Rodrigues, What is Natural Language Processing, 2017, Available at:
<https://medium.com/botsbrasil/o-que-%C3%A9-o-processamento-de-linguagem-nat ural-49ece9371cff>

Jones Granatyr, Introduction to Natural Language Processing, 2020. Available at:
<https://iaexpert.academy/2020/12/15/introducao-a-processamento-de-linguagem-na tural/?utm_source=rss&utm_medium=rss&utm_campaign=introducao-a-processame nto-de-linguagem-natural&doing_wp_cron=1677801456.9338579177856445312500 >Leandro Kovacs, What is natural language processing? [NLP], 2021. Available at:
<https://tecnoblog.net/responde/o-que-e-processamento-de-linguagem-natural-nlp/>

PrograMaria, Careers in Computer Vision: how to prepare to work in the field, 2020. Available at:
<https://www.programaria.org/carreira-em-visao-computacional-como-se-preparar-pa ra-work-in-the-area/>

OpenCV, documentation, 2023, Available at: <https://docs.opencv.org/4.x/> Adrian Rosebrock, OpenCV Thresholding (cv2.threshold), 2021. Available at:
<https://pyimagesearch.com/2021/04/28/opencv-thresholding-cv2-threshold/>

Arnaldo Gualberto, Detecting objects with classical methods - OpenCV (cascades), 2018. Available at:
<https://medium.com/ensina-ai/detectando-objetos-com-m%C3%A9todos-cl%C3%A 1ssicos-opencv-cascades-440e29913b1b>

Adrian Rosebrock, pyimagesearch, OpenCV Face detection with Haar cascades, 2021. Available at:
<https://pyimagesearch.com/2021/04/05/opencv-face-detection-with-haar-cascades/ >

Jobu,YOLOv5 How to train a personalised object detection model, 2023. Available at:
<https://jobu.com.br/2021/06/28/como-treinar-um-yolov5/>

HSANKESARA, Flickr Image dataset, 2018. Available at:
<https://www.kaggle.com/datasets/hsankesara/flickr-image-dataset>

edureka, How to Implement Optical Character Recognition in Python, 2021. Available at:
<https://www.edureka.co/blog/optical-character-recognition-in-python/>
mordorintelligence, TRENDS, COVID-19 IMPACT AND FORECASTS (2023-2028), 2023, Available at:
<https://www.mordorintelligence.com/pt/industry-reports/computer-vision-market>
Alan Hevner et al, Design Science in Information Systems Research, 2004. Available at:
<https://www.researchgate.net/publication/201168946_Design_Science_in_Informati on_Systems_Research>
geeksforgeeks, Convert Text to Speech in Python, 2023. Available at:
<https://www.geeksforgeeks.org/convert-text-speech-python/>
geeksforgeeks, Pedestrian Detection using OpenCV-Python, 2023. Available at:
<https://www.geeksforgeeks.org/pedestrian-detection-using-opencv-python/> safe park signalling, 2023. Available at: <https://safeparksinalizacao.com/> freepik, 2023. Available at: <https://br.freepik.com/fotos-vetores-gratis/braille>

APPENDIX

GuiaInteligente" project, available in the public access repository at the following link:
https://github.com/williams181/GuiaInteligente

I want morebooks!

Buy your books fast and straightforward online - at one of world's fastest growing online book stores! Environmentally sound due to Print-on-Demand technologies.

Buy your books online at
www.morebooks.shop

Kaufen Sie Ihre Bücher schnell und unkompliziert online – auf einer der am schnellsten wachsenden Buchhandelsplattformen weltweit! Dank Print-On-Demand umwelt- und ressourcenschonend produziert.

Bücher schneller online kaufen
www.morebooks.shop

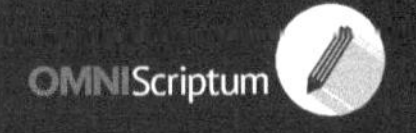

Printed by Books on Demand GmbH, Norderstedt / Germany